Creating Chaos & Mayhem
The Ultimate Guide to Disaster Exercise Planning

By Darryl Culley

Published by

Emergency Management & Training Inc.

65 Cedar Pointe Dr., Suite 144

Barrie ON Canada L4N 9R3

1-888-421-0665

www.emergencymgt.com

Acknowledgements

Creating exercises across North America requires a special team to ensure the highest quality, coordinated, and beneficial events. Putting the processes into a book also requires a special team.

A special thanks goes to Katherine Dale, Administrative Manager of Emergency Management & Training Inc. for her tireless dedicated work on all of our projects, and for marketing, coordinating our office and assisting with this book.

I would also like to thank Laura Kendall of Experts Writing Academy (www.expertswritingacademy.com), a fellow paramedic and crew member at TR events, for her encouragement, support and for at times kicking my butt to get this done on a short timeline.

Thank you to my beautiful wife Cheryl for all of her ongoing support.

Thank you!

Table of Contents

Prologue

A tornado ripped through the city on a Wednesday afternoon destroying houses, blowing out the windows of an elementary school and taking the roof off of a church which housed a day care program. It was June and nearing the end of "tornado season" but, in Texas, the risk was always there.

The weather forecast had predicted the potential for severe thunderstorms, hail and tornados, but it wasn't until the sirens sounded four minutes earlier that the threat seemed real. School children dutifully lined up in school hallways away from the windows while anxious eyes in the neighbourhood scanned the sky for the tell-tale funnel.

Then it was upon them, with the roar of a freight train, crashing through the western edge of the city.
The sirens of fire trucks, ambulances, and police cruisers filled the air. Public works staff quickly hooked up trailers stocked with barricades and headed out to close roads. At the same time, key city staff started reporting to the Emergency Operations Center to coordinate the response.

Although the teams took on their tasks calmly and with efficiency, following well-designed procedures, there was still an anxiety in the air.

Challenges arose that were not in the plans; assistance from other communities seemed to take forever to arrive, and parents were frantic for information on the safety of their children.

Several hours later the tension was broken with a well-timed meal and congratulations for a job well done. The next couple of hours would be spent identifying what could be done better next time.

Why next time?

Because this had just been an exercise. What if this experience had actually been real? Well, the city staff and responders had taken a huge step forward in being prepared and now had a list of tasks that they could do to be even more prepared for when disaster really struck.

Terminology

Throughout this book we will be discussing emergency and disaster exercises. For the purposes of this document we will use the terms "emergency" and "disaster" interchangeably. It will be your drill or exercise that will determine if the scenario is an emergency or a larger scale disaster. What you call it is really insignificant; what counts is what you are able to accomplish through the process.

Further, we will be using the terms "exercise" and "drill" interchangeably as well. While the term "drill" often refers to smaller incidents, and is used as a specific type of exercise, and "exercises" often refer to larger multi-agency incidents, we will use both terms throughout the book.

Some will insist that they are very different terms, however, using the same term repeatedly results in some words being used several times in a paragraph or sentence so we have decided to use the words interchangeably in most of the book.

Introduction

You're probably wondering if you really need drills/disaster exercises.

The answer is: that depends!

Do you want to have a realistic assessment of your policies, procedures, training, communications, etc.? If the answer is yes, then a well developed and implemented exercise is the closest resemblance to a real incident and can be a very effective tool to conduct that assessment.

In emergency services, emergency management and business continuity/continuity of operations, it is widely accepted that unless you can demonstrate your ability in a practical sense, there is no real assurance that you will be able to effectively respond in a crisis.

An outstanding exercise is not thrown together in a day. An exercise done well takes time, commitment, teamwork, and many months, or even a year, are required. This book will assist by walking you through the key steps to develop a successful exercise.

Participants in disaster exercises often look forward to the opportunity to work with a multitude of other agencies to solve a potential crisis, while others see it as an intimidating test of their role in front of a large crowd.

A quick social media survey revealed that the biggest impediments to getting authorization to conduct an exercise were:

#1 - Inadequate funding. By far most common reason for not conducting emergency exercises or to limit exercises to discussion based exercises is the lack of funding. The lack of funding was also a reason given for lack of preparedness activities including prevention and mitigation strategies, slow policy and procedure development (e.g. lack of staff to develop the policies and procedures), and lack of training.

One major hospital I have dealt with has a $750 million budget (yes, three quarters of a billion dollars) but only one emergency management person with a budget (besides their salary) of $10,000 a year to provide training and conduct exercises. All equipment, software and other tools have to be requested item by item.

One thing we have learned over years of experience is that just because an organization is large and looks well prepared, doesn't mean it is.

#2 - "We'll know what to do" and *"It seems like common sense"* are common excuses for not conducting drills and it is often the same excuse for not investing in the Emergency Plan program.

#3 - "It has never happened here" and all the variations of that comment are familiar explanations for not authorizing the resources required for exercises.

#4 - "You can't be prepared for a disaster anyhow" or "We could train for one thing and another could happen" are also common responses.

Unfortunately, it often takes a crisis that results in injury or death before these objections are set aside and increases in funding suddenly become available.

To assist in overcoming these objections, we have found that including senior managers and politicians in a well planned and executed exercise (as either a participant or observer) can result in a greater understanding and willingness to support emergency management programs. Even if the exercise is not of the size and complexity that is desired, they can be used to move the program forward.

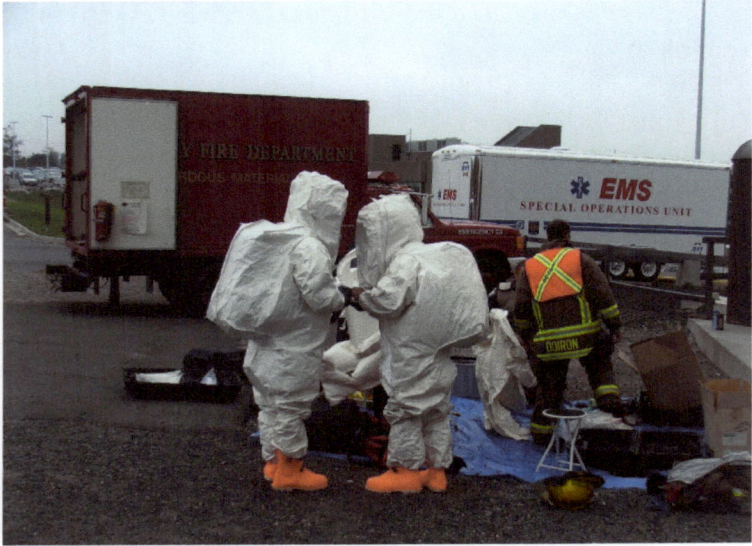

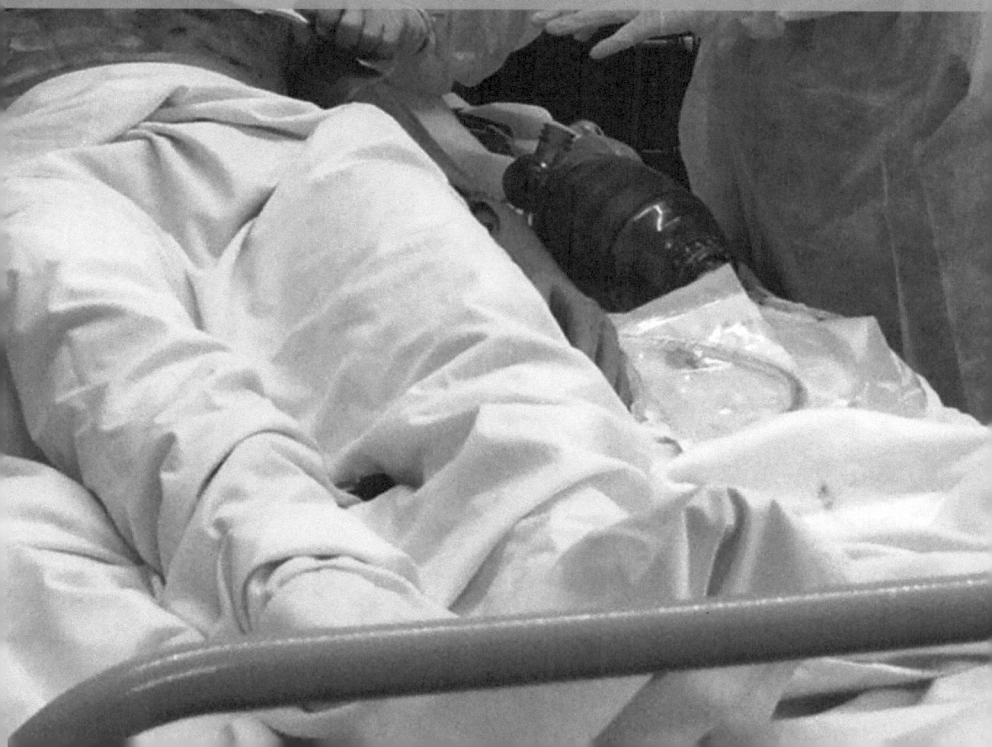

UNIT A
BED
1

CHAPTER 1
BENEFITS OF DISASTER EXERCISES

Often there are challenges getting the proper supports to implement a large scale exercise or to get participants interested in being a part of the event.

There are many benefits to a disaster drill but they are rarely communicated thoroughly. Properly articulating the benefits can result in larger budgets, more senior levels of interest, and greater participation.

The benefits to disaster drills and exercises include:

Staff Training

A drill is a practical training tool where the theoretical, class training and smaller hands on training is combined into a realistic scenario. The repetition of earlier training is now combined with team work, and the pressure of realism to reinforce the teachings. The lessons learned in an exercise will stay with the participants for a long time and be at the forefront of their minds.

For one of our clients we conducted an EOC (Emergency Operations Center) exercise involving an aircraft down in a remote northern community. The following year we followed up with a second aircraft down exercise that combined both the EOC being operational at the headquarters and a functional exercise at the same time, more than 1,000 kilometres away. Unfortunately, two years later one of their aircraft actually went down near the remote northern community that we used in the first scenario, killing all on board. The lessons learned in the exercises were directly applied on the day of the tragic crash.

People will typically favor one of the three primary learning styles (Visual, Auditory and Kinesthetic), so exercises are ideal learning opportunities since exercises usually provide all three when done right.

Test Plans and Procedures

Exercises can be used to test the plans and procedures that have been developed for your organization. Plans and procedures are usually developed by well qualified personnel, but it can only be projected that they will work until they are tested by an incident (exercise or real). Drills give opportunity to see how well the plan works and provide for opportunity to improve the plan before being needed in a real incident.

Identify Planning and Resource Gaps

When it comes to complex incidents it can be difficult to determine, on paper, where there are planning and resource gaps for the response. One may feel that "everything" has been covered only to discover, through drills, that items, issues, and needs have been overlooked.

Test Mission Critical Assumptions

A thesis or assumption is just that until it has been tested. An exercise gives the opportunity to test a hypothesis in a hands-on setting.

Building Working Partnerships

When we have opportunities to work with others we create friendships and partnerships. We learn about each other's skills, talents, and knowledge. We get to help each other out and create bonds.

It has been said that it is easier to get angry and frustrated with a stranger than a friend. We tend to put more efforts into the mutual partnership when we know each other.

Exercises have the benefit of bringing people together.

Test Communications and Interactions Between Agencies

In major incidents, often the biggest challenge is communications. Communication breakdowns come in a number of areas including technology, message transmittal and understanding, sorting through volumes of messages for critical information and people feeling out of the loop.

Practice makes perfect...or at least much better. The best opportunity to practice communication is during a drill where the stresses and realism are similar to an actual incident.

Test Technology

The introduction of technology to crisis response and management is designed to assist the teams in working more efficiently and effectively. Until the technology is tried in a real incident it is unknown how your team will use and interact with technology. The technology may be able to do hundreds of tasks but if the users do not feel comfortable using it or go back to old processes, it is worthless.

Testing and practicing the technology in a simulated environment will provide insight into its effectiveness, provide opportunity to identify enhancements, and demonstrate the benefits that it can provide. The more realistic the scenario, the more effective the test will be.

Accreditation, Legislation and Contractual Requirements

Many jurisdictions have mandated requirements for drills and exercises. Whether it is a requirement for fire drills, evacuation exercises, active violence drills, or other mandated requirements, legislated requirements may specify the type and number of drills that take place.

Organizations that are accredited such as hospitals and long term care facilities, or organizations holding or working towards international standards recognition, may be required to hold drills, tests and exercises to demonstrate their ability to meet the accreditation standards.

At times, there are contractual requirements that an organization must demonstrate its ability to respond to crisis. This is often required in business continuity where parts suppliers are required to demonstrate their ability to maintain continuity of supply to the end user, particularly in a "just in time" manufacturing environment.

Providing Reassurance to Clients and Stakeholders

While providing reassurance to clients and stakeholders is rarely a driver for holding a complex exercise, if marketed appropriately, it can be used as a tool to connect.

It is extremely important for parents to know that their children are well protected at school, or for children to know that their parents and grandparents are well protected at nursing homes. Community members want to know that their emergency services are trained, capable and ready to respond when a disaster strikes. An exercise can be marketed to the targeted stakeholders as a demonstration of your readiness to protect them.

Deterrent

At times, exercises are used as a deterrent to those who may be considering an act of ill intent. Police forces, security services, special response teams and the military may use exercises to demonstrate their abilities to rapidly respond to any act of aggression to deter potential threats. When a criminal or organization sees the overwhelming response they may face, it may dissuade their intent.

Such displays of responder skills, resources, and strength are often made public before major sporting events and high profile VIP gatherings (e.g. gathering of international leaders).

Building Problem Solving Skills

The best way to build problem solving skills is to practice them in various scenarios. A well designed exercise can create various challenges for the teams to resolve and evaluate to provide additional options to improve performance.

Demonstration to Funding Sources

Exercises have the ability to demonstrate to funding organizations the value of the services provided as well as the benefit of doing the exercises as a training platform. Emergency managers often complain about the lack of funding or priority placed on their programs. Exercises are opportunities to demonstrate to the funding and approval sources that what they do is a vital component to the organization.

Funding is critical for any emergency preparedness program. Whether the program is operating on a shoestring or has a well-financed base, it is important to demonstrate how effectively the money has been spent and the importance for continued or enhanced funding for preparedness activities including training and drills.

Set Goals for Future Exercises

The conclusion of an exercise is not the conclusion of the preparedness process. In the after action report there will be a series of accomplished items, successes, recommendations and opportunities for improvement.

The exercise assists in building the next steps of preparedness, including how to build on the drill to set the goals and objectives for the next exercise.

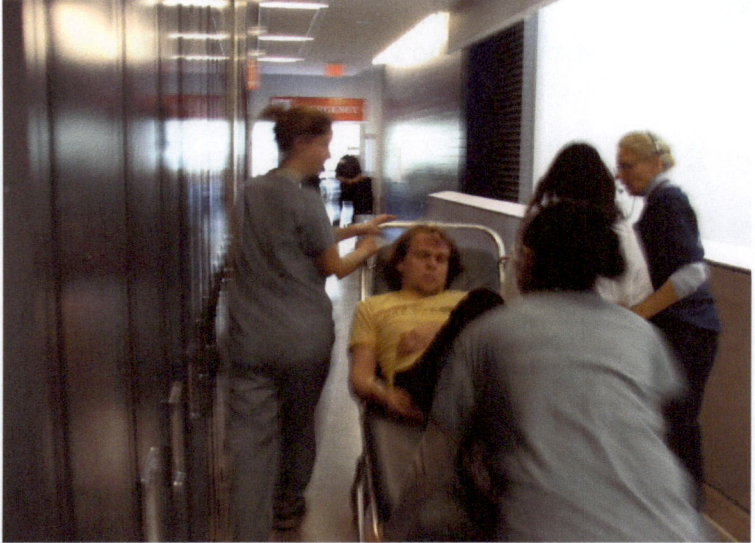

CHAPTER 2
GOALS, OBJECTIVES AND APPROVALS

The earliest steps of the exercise process include establishing your primary goals and objectives, and securing the approvals required to conduct the exercise.

Exercise Goals

The exercise goals establish the overall aims of the exercise and provide direction to keep the planning and scenario in line with the intent of the exercise.

An exercise may have more than one goal, but in order to keep the event manageable it is recommended that the exercise be limited to three or four goals.

Goals are the broad, over reaching aims. Examples of realistic goals are:

- Test the emergency plan for the ability to facilitate the rapid evacuation of _____

- Test the working relationship between the university staff and the city emergency responders.

- Evaluate the communication strategies and technologies between the emergency operations center and the public.

- Determine the ability to restore critical services to
_____

Establishing Realistic Objectives

Once the goals are established, objectives for each of those goals will be developed. For each of the goals there will be three or four objectives.

When selecting the objectives, consider the key benefits you want to achieve. What do you want to learn?

Keep the objectives simple, achievable, and measurable.

Asking "what will success look like when the exercise is complete?" will assist in identifying some of the objectives.

If the exercise is a multi-agency event, keep the objectives limited to three or four per agency in order to avoid placing too many expectations on a single organization or making interactions between the agencies too complex.

Here are a list of potential areas that can be used in developing your goals and objectives:

- Testing internal communications
- Testing communications technology
- Testing external (stakeholder or public) communications (including social media)
- Information gathering and recognition of warning indicators
- Analysis of information gathering
- CBRNE (chemical, biological, radiological, nuclear, explosive) detection
- Protection of critical infrastructure
- Restoration of critical services/functions provided by the organization
- Epidemiological surveillance and investigation

- On-site incident management
- Emergency Operations Center management
- Responder health and safety
- Volunteer management and donation management
- Hazardous materials/CBRNE response and decontamination
- Evacuation and/or shelter in place
- Shelter operations
- Search and Rescue
- Emergency triage and pre-hospital treatment
- Medical surge (code Orange)
- Fatality management
- Damage assessments, and
- Restoration of critical infrastructure

Approvals

Exercises can grow large and complex, involving staff time, technological and equipment resources, and may have far reaching implications involving policy revisions, public/media attention, and exposing gaps and vulnerabilities in the emergency preparedness plan.

With these investments and potential risks it is important to obtain the appropriate approvals.

Who do you want approval from?

There may be people within the organization that are not required on the official authorization, but you would still want their approval. For example, your exercise may have all the approvals required at a deputy chief or director level, but having the approval of the chief and CAO or COO may provide a higher profile for your event and gain additional support from other departments or agencies at a higher level.

Obtaining the highest level of approvals on the exercise will not only bring important cooperation and an increased profile, but it may also increase the scrutiny of the drill as there will be more senior levels in the organization watching. Getting a higher level of approval for the exercise will make it even more important to be vigilant to ensure a successful exercise.

CHAPTER 3
EXERCISE FORMATS

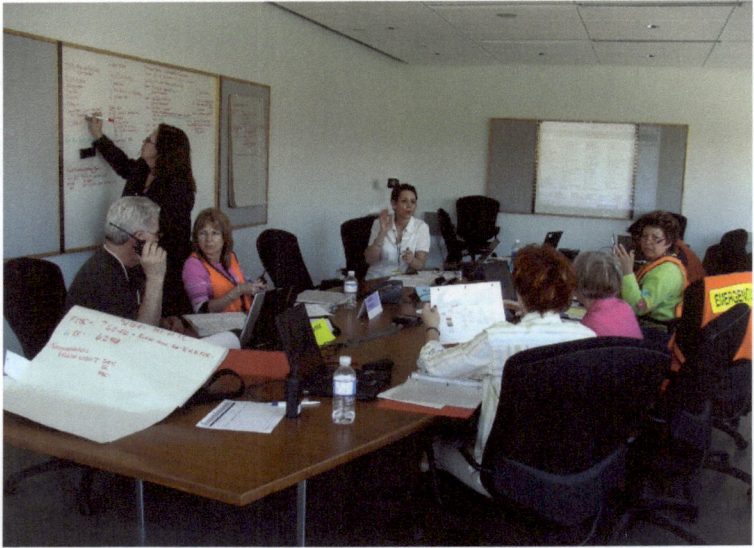

Discussion-Based Exercises

- Seminars
- Workshops
- Tabletop Exercises (TTX)
- Games

Discussion-based exercises stimulate discussion around hypothetical situations, review plans and procedures, strengthen problem solving skills, and educate staff and managers. These exercises are generally low cost and low risk in comparison to operational exercises.

Discussion-based exercises include the following:

Seminars

A seminar is an informal discussion led by a facilitator. Seminars are designed to provide an overview of the plans, policies, procedures, strategies, resources available, etc. They can be a good starting point for introducing new or revised plans and can be used to gain input on the emergency plan development (e.g. a seminar to review new evacuation procedures).

Workshops

A workshop resembles a seminar, but is focused on developing a specific plan or procedure. Workshops have a higher participant interaction and involvement than seminars.

To be effective, workshops need to be directed towards a specific issue and an outcome or goal has to be clearly defined (e.g. a workshop to develop mutual aid agreements or create a new standard operating procedure).

Tabletop Exercises (TTX)

A tabletop exercise involves key personnel discussing simulated scenarios using a controller to facilitate the scenario. Tabletop exercises can be used to assess plans, policies, and procedures for a defined incident. They are typically aimed at facilitating an understanding of a process and/or identifying strengths and areas for improvement of a plan. For a tabletop exercise to be effective, the participants must be committed to the process.

Games

A game is a simulation of operations that often involves two or more teams, usually in a competitive environment, using rules, data, and procedure designed to depict an actual or assumed real-life situation. Because of the competitive nature of "games" it is often used to train in scenarios where there is an adversarial function (e.g. police vs. criminal activity, war games, etc.). Games are focused on the personnel and their ability to integrate existing plans and resources into problem solving scenarios. Games can be slow and methodical or rapid and stress inducing.

Operational Exercises

- Drills
- Functional Exercises (FE)
- Full-Scale Exercises (FSE)

Operational exercises are used to validate plans, policies, agreements and procedures and ensure that they are practical. These exercises are used to clarify roles and responsibilities, identify equipment and resource gaps, create stressors on the system to see how the system holds up, evaluate organizational response, and provide an opportunity for stakeholder observation of the response in motion.

Operational exercises are more complex than discussion-based programs and therefore have increased planning and costs.

Operations-based exercises include the following:

Drills

A drill is a coordinated, supervised activity usually employed to test a single, specific, operation or function within a single agency or organization (e.g. a fire department conducts a decontamination drill or a hospital conducts an evacuation of a wing).
Drills can be used to provide training on new equipment, practice and maintain current skills, and validate procedures.

Drills can also be used to determine whether procedures can be used as designed, assess whether additional training is required, and to reinforce best practices.

Functional Exercises (FE)

A functional exercise examines and/or validates the coordination, command, and control between various multi-agency coordination centers (e.g. emergency operation center, on scene command, etc.). A functional exercise does not involve any "boots on the ground" (i.e. first responders responding to an incident in real time).

Full-Scale Exercises (FSE)

A full-scale exercise is a multi-agency, multi-jurisdictional, multi-discipline exercise involving functional (e.g. joint field office, emergency operation centers, etc.) and "boots on the ground" response (e.g. firefighters decontaminating mock victims).

Photo: Large full-scale exercise

CHAPTER 4
BUDGET

Every drill has a cost attached. When designing the exercise one must consider the resources required for the size and complexity of the exercise and the associated finances available for the drill.

In some cases those costs are partially absorbed through the utilization of staff and resources that are already on duty or present. However, the salaries and benefits are costs that still need to be taken into consideration as these resources could be utilized on other projects.

Other exercises may require increased staffing levels, participants to be brought in on overtime or travel costs to bring teams together. They may also require specialized equipment such as aircraft, boats, and heavy rescue equipment.

Costs not to be overlooked include the time and associated costs of the planning and organizing of the exercise.

Costs to be considered include:

- Exercise planners (time, travel, expenses)
- Site costs
- Set up costs including site simulation
- Participant salaries
- Costs of bringing in extra staff
- Transportation of equipment
- Equipment maintenance costs including wear and tear

- Equipment operational costs such as fuel

- Replacement of lost or damaged equipment

- Disposables (firefighting foam, medical supplies, etc.)

- Equipment rental costs

- Volunteer costs (food, transportation, supervision, etc.)

- Swag (gifts given to participants and volunteers for participating such as t-shirts)

- Insurance

- Demobilization and clean up

- Moulage (makeup to simulate injuries)

- Special effects

- Props (old buses, vehicles, aircrafts)

- Security, and

- Consultant costs (may actually decrease overall planning costs)

Each of the above must be given some consideration. We will use volunteer costs as an example.

It may appear at the outset that there are not any costs attached to utilizing volunteers, and sometimes that may be the case, but often there are costs attached. There may be the requirement (or good will) to provide the volunteers with food and drinks, transportation, and small rewards/swag. Further, there may be the costs of paid staff recruiting, training/orienting, and supervising the volunteers.

One exercise at a health care facility utilized 30 grade 12 students to play the role of patients, but the school required that the health care facility cover the cost of renting a school bus to transport the "patients" back and forth to school. This facility also provided snacks and drinks for the students and the teachers as a thank you. One of the benefits was that the school provided the teachers to supervise the students at no cost to the facility.

We have used t-shirts as incentives for volunteers with "I survived exercise _____" printed on the back and our logo on the front. These turned out to be extremely popular when the students got back to school to tell their stories of disaster and how they were rescued. The t-shirts had a cost attached but served their purpose extremely well. For other exercises, we have had participant draws for gift cards to restaurants, electronic stores, etc.

Full scale exercises may require the rental of equipment including backhoes, generators, portable lighting, propane heaters, and even RVs, which we have used as our exercise control center/sleeping facilities for exercise staff.

Frequently, scrap vehicles may be donated for the exercise to be used as props, but the scrap yard or garage may request the cost of towing to and from the site be covered. This is more common when using larger vehicles such as buses, trucks and aircraft as the cost to transport these is higher.

Sometimes municipal rules add costs to the exercise. In one city exercise the police department had old cruisers that were stripped of their equipment ready to go to auction for disposal. These vehicles would be used in the exercise to be cut up for the extrication of patients. When the exercise planners, including the police department, requested two of these old cruisers for the exercise, they were invoiced by the finance department for the lost revenue from the auction. While this was frustrating to the exercise planners, the finance department had a protocol to follow. Therefore, it is important to ensure that all potential costs have been investigated and assessed.

Sometimes the costs come following the exercise as part of the demobilization. An organization that used portable shelters in their deployment found that if the shelters were not totally clean and dry when being packed away, the equipment would start to mold and deteriorate. Therefore, after a winter deployment or one in the rain, a warehouse had to be rented and staff paid to re-setup the shelters for cleaning and drying before being packed away awaiting deployment. This also allowed them to assess the condition of the equipment, do repairs as needed, and confirm that all of the equipment was accounted for and packed in the proper containers, which doesn't always happen during tear down following an exercise or actual deployment.

In one exercise, federal and provincial agencies used the same brand of shelter and connected them together during a joint exercise. During take down, the parts of the shelters were mixed and did not all reach their appropriate owners. One agency did not re-setup until a deployment months later only to find that key parts to their shelters were missing, presumably mixed up with the other agencies on the earlier drill.

One way to minimize costs is to use an experienced and professional exercise planning consultant. An experienced consultant can often come in and address the tasks in a reduced time frame due to their experience. Further, it reduces the workload on those who may already be overwhelmed with their day-to-day office activities.

An experienced consultant can often foresee challenges and avoid or mitigate risks that those less practiced would not be able to identify in advance.

A professional exercise consultant can often bring additional resources to the exercise including logistics, equipment, specialized staff for various roles (e.g. physician patient information providers), moulage and special effects and media etc.

An added advantage of a consultant is that once the exercise is over, you no longer have to carry the salary costs of an employee or employees.

A great consultant will make the client look good by giving them the credit for the successful exercise.

CHAPTER 5
PLANNING TEAM

The quality of your planning team will have a direct impact on the quality of your exercise. A planning team should include a small set of organizational leaders and exercise experienced personnel.

The planning team should be kept to those essential to the process, as the larger the "committee" the more cumbersome and time consuming the planning is. Keep in mind that the planning process is very time consuming to begin with and is only made more intensive as the complexity of the exercise and the number of participants increases.

A capable leader is required to keep the planning team and all of the others who wish to have input on track and focused.

Planning Team Format

The Incident Management System (IMS) is the internationally accepted organizational structure for responding to incidents of all scales and all types.

IMS can be used for both emergency and non-emergency events (e.g. the planning of a large public event or an exercise). The use of IMS in planning and implementing an exercise is consistent with the use of IMS in response and keeps the similar terminology and functions.

IMS is an expandable system based on functions not positions. Each function is assessed to see if it is required for the incident. A function may be fulfilled by one person or a team of people. For smaller events one person may fulfill multiple functions, whereas large events may have a team of people working on each function.

This is not intended to be an explanation of IMS, but to quickly demonstrate how IMS can be used by the design team for planning exercises.

Incident Manager

Overall responsibility to organize and direct the exercise design and implementation.

Operations

Operations is the function of carrying out the exercise. This would include the controllers and SimCell personnel.

Logistics

Logistics is the function of organizing and supplying staffing (e.g. volunteer coordination), equipment, maintaining the physical environment, food, water, and supplies to support the exercise.

Planning

The planning function develops the scenario and determines the implications of the scenario on the resources required to conduct the exercise. When using IMS for a multi-agency exercise, the planning function often has a representative from each of the main agencies participating.

Administration/Financial

The Administration/Financial function monitors the exercise costs, coordinates liability releases, ensures insurance coverage, provides administrative support to the senior IMS team members and ensures documentation of all meetings.

Public Information

The Public Information function organizes communications with the stakeholders, public notifications, and the media (as appropriate), coordinates marketing, and provides information updates throughout the exercise.

Liaison

Liaison is the function of communications to create cooperation between internal departments and external agencies. Liaison has the role of being the contact representative.

Safety

In every exercise the health and safety of participants is paramount. The safety function monitors and has authority over the safety of operations. This process includes conducting a HIRA (Hazard Identification and Risk Analysis) of the exercise, implementing prevention and mitigation strategies to address the potential hazards, and monitoring the exercise itself.

As the scale of the exercise dictates, each of the functions above may have an individual or team to assist in the meeting of their tasks.

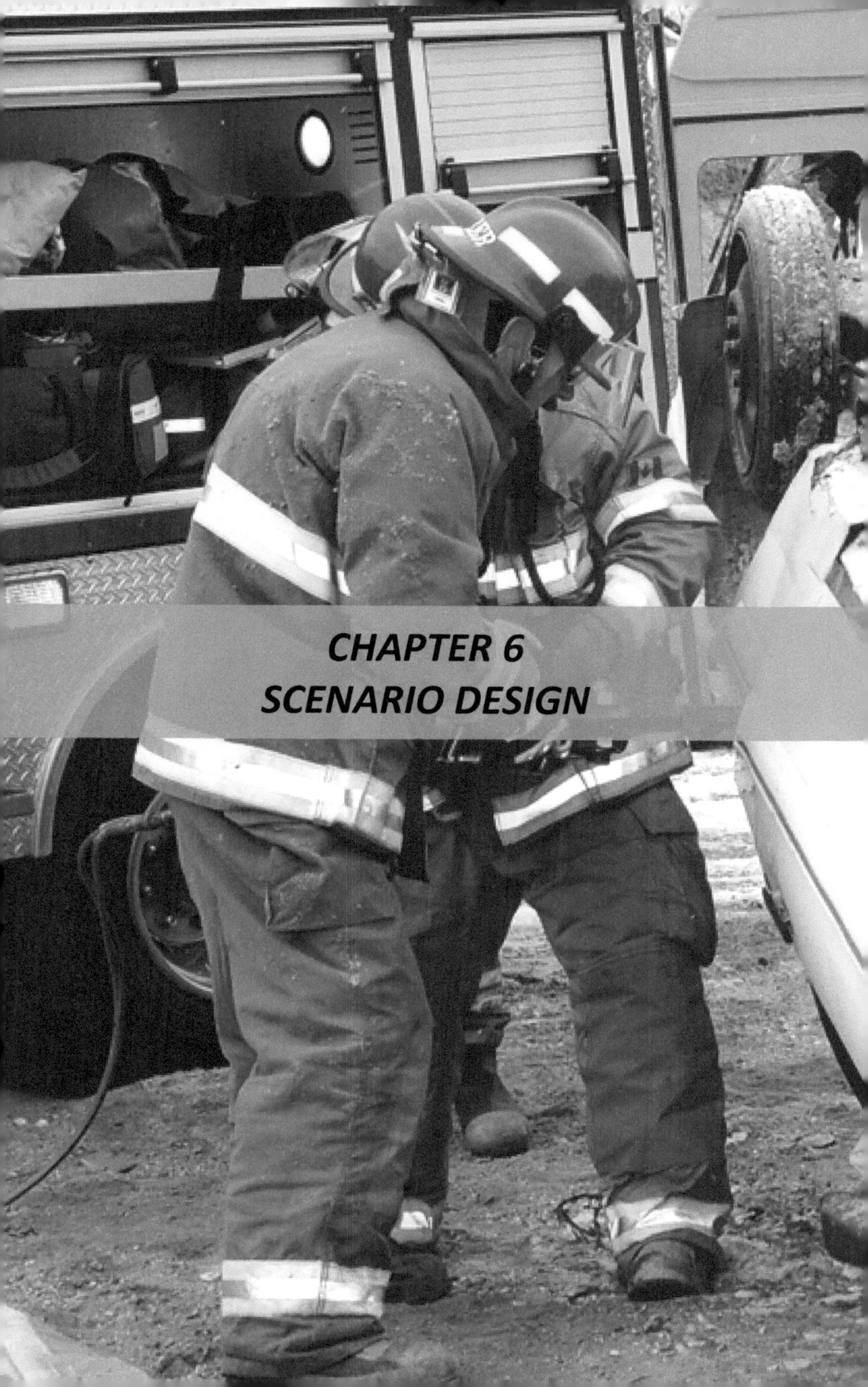

CHAPTER 6
SCENARIO DESIGN

Designing the scenario requires creating a description of the background information, current situation, and the inputs and injects that will drive the exercise players to take action. Injects and inputs are the messages and queues used to provide information to the exercise players.

It is important that the scenario design is based on your goals and objectives. The scenario must ensure the exercise is meeting the outcomes you have identified.

The exercise should be realistic to the threats and risks of the community. Using the local HIRA will ensure that the exercise is practical and realistic.

It is important that the scenario be believable, knowing at the same time that some creative license applies at times, in order to keep the exercise moving along.

On occasion it is easy to get carried away in the scenario design process. At one planning meeting I observed the scenario grow from a traffic collision between a bus and a truck to having the truck driver flee the scene to an apartment building where police officers entered in foot pursuit to find a meth lab which then exploded creating an officer down situation with an apartment building fire involving hazardous materials and a hundred building occupants. After a break the group refocused and started again with a simpler scenario.

It is important not to create a total catastrophe or apocalyptic scenario that is really no win. Often exercises keep ramping up the injects creating a more challenging scenario. In reality, the situation is often improving with the response, so realistic effects of the decisions made should show progress in resolving the situation.

When designing the scenario it is important that goals and objectives can also be met within the timeframe allocated. Although there may be some "time warp", the scenario should allow the participants to work through the goals and objectives within the time set out.

Time Warp

In an exercise, time becomes relative. For example, a good controller will speed up or slow down injects and timelines based on the capacity of the participants to deal with the exercise.

Discussion-based and EOC exercises often will proceed faster than they would in reality. Decisions are often made quicker, having less information and fewer external inputs than would occur in reality. Further, in a real event the emergency response component may take many hours or even days before other components of the response are put into place. Therefore, to exercise many of the additional components a time warp may be required.

Time warps can be accomplished in a couple of different ways. The first is to use a clock that is set to run at 1.5 or 2 times the actual speed. The second is to suspend all time based reality and to move through the stages or overlap the stages of the response in a more rapid fashion.

For example, in a tornado scenario we conducted in Dallas, Texas, the teams responsible for windshield assessment teams for the public works and utilities would not be sent out until the emergency services response had concluded the rescue component and the roads were clear for the teams to drive through. However, in a half day EOC exercise the concept of time had to be suspended to get these components included in the exercise.

The determination of the timeline for the exercise will be dependent on the goals and outcomes of the exercise.

In one exercise involving a "car bomb" scenario at a hotel that was being torn down, the decision was made to have a real time 72 hour scenario for the response teams. It took more than 24 hours for heavy urban search and rescue teams to arrive from across the country, set up their equipment, assess the site, start the search with K9 units and then move into the rescue of the "trapped".

But not every agency has the ability to have dedicated teams assigned to a site for 72 hours taking shifts, therefore, some of the tasks can be time warped in order to meet time limitations.

Exercise Examples

There are thousands of scenarios that can be developed into great exercises. Many of the scenarios can be linked or combined.

Some of the examples that could be used include:

- Aerosol Anthrax in a transit system, office building or shopping mall

- Pandemic
- Chemical spill (manufacturing or storage facility, rail, marine, trucking, etc.)
- Chlorine tank explosion (e.g. train derailment or at a swimming pool)
- Propane tank/truck explosion
- Earthquake
- Tornado
- Hurricane
- Power failures
- Ice storm/snow storm
- Improvised Explosive Device
- Food contamination
- Cyber Attack
- Flooding
- Pipeline failure (e.g. natural gas, oil, water, steam)
- Train derailment/collision (freight, passenger, or both)
- Cruise ship / Ferry (fire, disease outbreak, collision, storm)
- Aircraft crash
- Motor vehicle collision (bus, truck, automobiles)
- Wildfire

- Structure fire/explosion, and

- Illicit drug lab explosion

- Zombie apocalypse / Space alien invasion

Inject Examples

Some of the inject examples below are used within the exercise to challenge the "expected" with experienced responders that are already capable of dealing with the standard incidents.

Here are a number of scenarios that can be added to an exercise to create challenges and interactions. They will require additional planning, coordination, and implementation but will also create opportunities to test decision making, collaboration between agencies, and the flexibility of the responders.

Officer Down

During the rescue scenario a responder (pre-identified) will go down with a serious medical condition or injury at the height of a rescue.

K9 Down

We have used this scenario a number of times in working with hospital emergency departments and mobile hospitals/treatment centers. During a medical response scenario we would introduce a K9 handler (police or rescue) who carries in his canine partner advising the dog has been injured during the crisis. The handler demands that the hospital/triage center provide care for her/his partner.

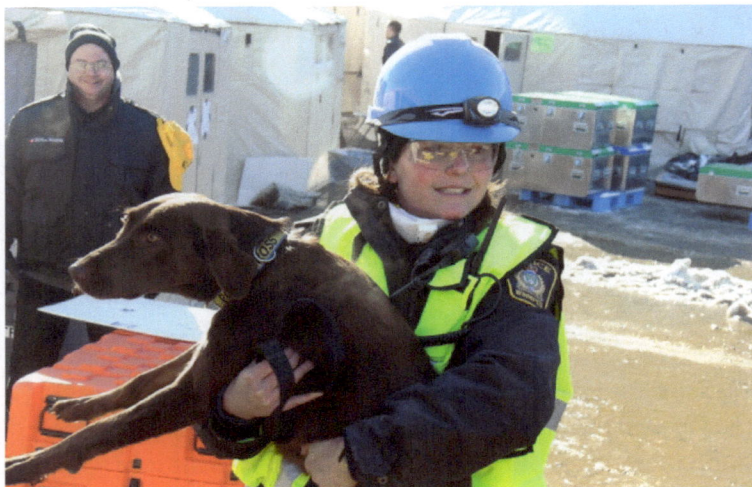

Santa Claus

One scenario we planned was a Christmas Parade through a downtown area that was next to a railway line. During the parade a train derailed causing a release of a toxic gas resulting in patients ranging from parade watchers to many dressed in costume including Santa Claus.

To expand this scenario to include other agencies such as the SPCA, the parade included horses in the parade that became ill as well as others that were spooked lost their handlers and raced through the crowd causing human injuries and creating a containment and capture issue.

Politician

There is always fun using injects with politicians. The following are a couple of sample injects:

- A senior politician/entourage arrive at the scene or hospital wanting to speak to patients, get a photo opportunity, or to demand special/prioritized care for a relative.

- A high level politician arriving at the EOC creating political interference is used to evaluate if the pressure will impair or change the decision making of the Incident Command team.

Animal Shelter Explosion

A tough scenario for animal lovers, but one that creates ethical discussions is the animal shelter explosion. In this scenario a combination of human and animal victims are mixed in the response area. Using animal manikins for severely injured or dead animals adds to the setting.

Prisoner Transport

In one scenario we simulated a transport carrying a HazMat load travelling down a highway exit ramp and crashing into a number of vehicles at an intersection at the end of the ramp. One of the vehicles was a prisoner van with the guards being severely injured and a mix of injuries obtained by the "prisoners" who were handcuffed and shackled.

The scenario included one prisoner attempting an escape at the scene and others being extremely uncooperative at the scene and in the hospital emergency department. The additional chaos, in addition to the mass influx of patients, pushed the limits of some staff. Keep in mind that as a HazMat call, all of the patients had to be decontaminated by the emergency services.

Flood – Business Continuity

For many facilities (hospitals, long term care/nursing homes, corporate buildings etc.) the infrastructure is located in the basement and the ground floor levels. The infrastructure often includes computer networks, electrical rooms, elevator service rooms, etc. and many include other areas such as document storage. A flood can create numerous business continuity issues.

For the scenario, take masking/painters tape and mark a line along the walls of the basement and/or ground floor to indicate the water level. The line will give a visual representation as to the equipment and items that would be damaged by the water level.

CHAPTER 7
SITE CONSIDERATIONS

Exercises may be held at various locations depending on the type and extent of the drill.

An EOC exercise may be held in the designated EOC or the scenario may dictate that the EOC has been destroyed and the team must function from a backup, mobile or impromptu location.

Functional exercises may take place within a building, facility, training center or campus of the organization. Alternatively it may occur off-site in a park, quarry, parking lot, intersection, etc.

The following are examples of where exercises have been conducted:

- Hospital under construction - simulated the new emergency department

- Closed hospital wing at a psychiatric hospital - simulated an emergency department

- Hotel undergoing demolition – simulated a car bomb in the lobby

- Quarry – simulated a motor vehicle collision on the highway

- Fire department training facility – simulated a motor vehicle collision on the highway

- Closed hospital – simulated an active gunman in a school

- Closed rehabilitation center – simulated public protests and riots

- Active airports – simulated a plane crash

- Lakes and rivers – ice water and swift water rescues

- Parking lots – simulated intersections

- Industrial facilities

- Rail yard

- Transit system station, and

- Sports stadiums and arenas

The space being used must be large enough to accommodate the exercise. Space considerations include:

- Exercise space requirements for the incident and response agencies

- Set up space/rooms (e.g. for moulage and makeup)

- SimCell room(s)

- Participant and volunteer parking

- Washroom facilities

- Climate control for registration/preparation areas

- Meeting and food locations

- VIP and Media area and parking/access

- Debriefing (often referred to as a "hot wash") location, and

- Shelter in the event of a severe storm

It is critical to assess the security of the exercise site and the ability to limit public access to the exercise area, protect expensive equipment and ensure safety when heavy equipment, aircraft, or tactical maneuvers are taking place.

The ability of the public to view the exercise creates potential issues. Exercises can create traffic congestion and dangerous situations when drivers slow or stop to observe. The public will often want to take photos, video and watch the action. Further, if they do not understand that it is an exercise they may call 9-1-1, try to assist in rescues, and/or attempt to gain access to the site.

While buildings that are not being used may be attractive for an exercise, there are a number of factors that need to be considered when using sites under construction that have been closed or are being demolished.

The factors to consider include:

- Safety of the participants due to the structural integrity of the building, heavy equipment being present, debris/construction material present or potential safety barriers/equipment not in place yet or missing

- Construction or other workers on site

- Insurance coverage

- Legal ownership of the site and appropriate authorizations to use the site

- Lack of functioning services (e.g. washrooms, electricity, heat, etc.), and

- Life safety systems being activated due to the scenario (e.g. fire alarms, sprinklers)

Other considerations include the suitability of the site for the type of response. If tents need to be set up using stakes, a newly paved parking lot may not be appropriate. Other considerations would include evaluating where water collects during heavy rain.

Whenever an exercise is being planned a Hazard Identification and Risk Analysis of the site will assist in determining what the potential hazards and risks are.

CHAPTER 8
INTEGRAL COMPONENTS

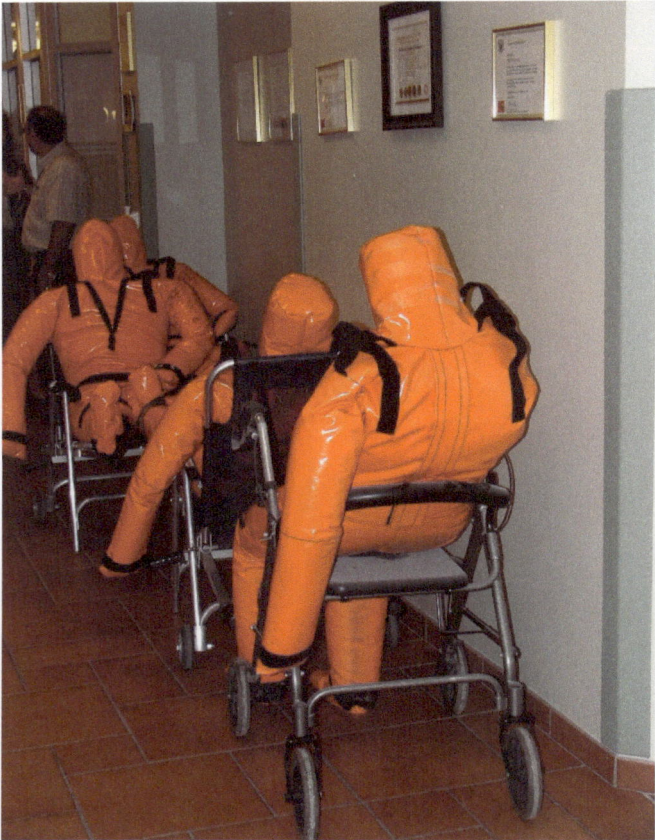

Pre-Training

While the exercise is a training opportunity, it is important to ensure that it is successful. The successful implementation of an exercise starts well before the date of the exercise.

If you overextend and stress the people participating beyond their capabilities, there will be resistance to future exercises and events. People often do not absorb as much information when they are frustrated or embarrassed.

People will learn more if they feel that they have been successful. They will understand that some things could have been done better and will identify how they can improve for future exercises and actual events. Therefore, it is important to ensure that the participants have been properly trained in the roles and responsibilities that they will be undertaking in the drill.

For larger or more complex exercises, much of the lead time for planning can be simultaneously utilized to provide training for the participants. The training may include smaller and less complex drills, table top discussions, and reviews of procedures and communications before leading to the large exercise.

Master Event List

The Master Event List outlines the major "events" or components of the scenario plotted out in a timeline.

Inputs/Injects

Injects and inputs are the messages and queues used to provide information to the exercise players. They can be in the form of pre-scripted conversations, notifications, or messages that simulate an activity or occurrence in the disaster. They are designed to provide information to the participants, create opportunities for response assessment, and to prompt participants to make decisions and take action.

At times the decision may be to ignore the input or sort through the inputs to determine which ones are actionable and which ones are irrelevant or superfluous.

An input or inject often originates in the "SimCell" (simulation cell/room) where exercise team members simulate information coming in to the drill participants. For example, in a tabletop exercise, exercise controllers may play the role of onsite police/fire/EMS officers and use a radio or telephone to call into the EOC.

The input and injects should reflect the level of responsibility that the participants would normally have. Front line staff should not be given decisions that would be several levels of decision above their authority, nor should Chiefs be focused on front line tactical decisions that would be made daily by their front line staff.

If the injects become too complicated, the team may begin to focus on the minute details and forget to work as a team.

Script

At times, controllers will want to cause a specific action within the exercise. To cause that action, a "script" is used to either prompt or advise that the action has been taken or activated. Scripts can be identified word-for-word or the SimCell personnel can be given the gist of the conversation, allowing them to ad-lib.

Ad-libbing can be done successfully when using experienced personnel in the SimCell but caution needs to be used when using relatively inexperienced personnel or those not familiar with the specific role. In these cases, more specific or word-for-word scripts are advised to ensure that the inject is transmitted and received as intended.

Often, when using experienced SimCell personnel we will use word-for-word scripts in specific and critical injects and provide general conversation points, allowing for ad-libbing in other areas, such as public calls or media calls.

Controllers

The role of the controller is to enable and guide the exercise and participants.

The controller enables the exercise by providing "inputs" and scenario "messages". These inputs and messages are to be received by the participants for information and to be actioned. They simulate information being received, visualized, read, heard, smelled, etc. The participants are to take this "input" or "message" and to go through the steps that would be taken based on that information.

The controllers guide through:

- Monitoring the actions and team processes within the exercise

- Speeding up or slowing down the "inputs" and scenario "messages"

- Monitoring for safety and well-being of participants (along with agency safety officers)

- Providing guidance or encouragement where needed

- Encouraging communications between the site and EOC

- Encouraging communications

- Setting time guidelines

- Encouraging actions if they have been missed

- Evaluating the exercise, and

- Calling a termination to the exercise

Controllers need to have both experience in exercises as well as with the setting in which the exercise is taking place. For example, a controller in an EOC should have

EOC experience, while a "site" controller should be an experienced responder.

Evaluators

Evaluators are often seen as "examiners" who are at the drill to pass or fail the participants. Our view of the evaluator is more of an empowerment facilitator, present to provide ideas to move the program forward.

An evaluator should be a subject matter expert and it is best if they have experience in a similar organization or similar type of emergency. Selecting peers from similar organizations or experts in the field to play the role of evaluators is a great opportunity to share experience and expertise.

Often controllers will double as evaluators; however, due to their focus on ensuring a smooth and effective exercise, they may miss some of the key points unless they are very experienced in conducting the drills.

Evaluators will assess the ability of the team and provide feedback and constructive criticism on the ability of the organization to meet the goals and objectives of the exercise.

A good evaluator will be able to provide constructive criticism and be able to identify the root concerns without being critical of individuals.

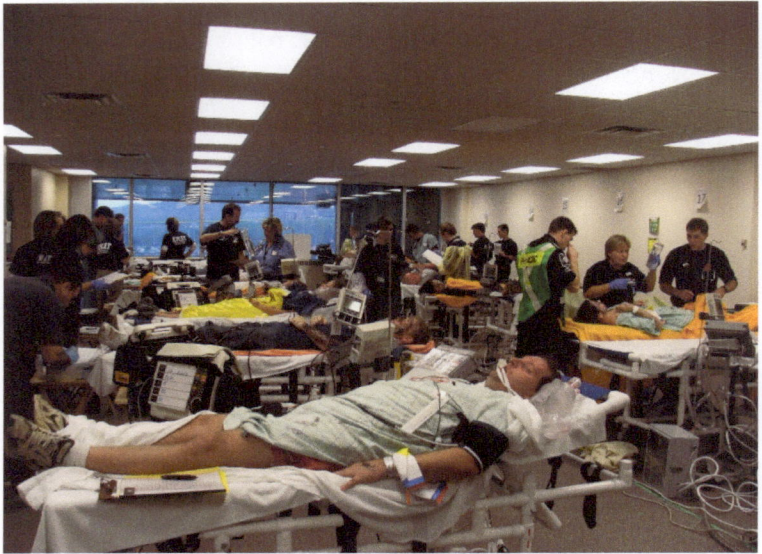

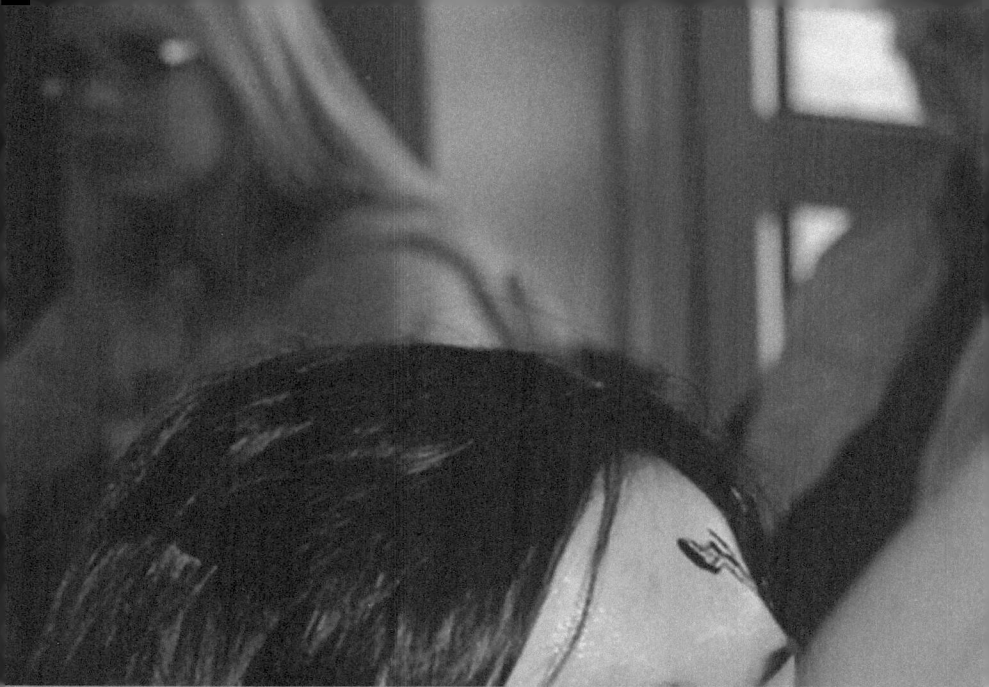

**CHAPTER 9
PARTICIPANTS**

Depending on the size and complexity of the drill, the number of participants may range from a handful to hundreds or even thousands. It is important in the design of the exercise to understand who the participating organizations are and each of their learning objectives.

Determine if the participants will be:

- Internal vs. external to your organization
- Involved in planning and coordinating or an actual exercise participant
- Participants or playing the roles of casualties and victims

It is important to have an active and meaningful role for each of the participants in the exercise. Having participants who have dedicated their time to the exercise and end up in a role which has nothing to do or has little practical value creates frustration and resentment.

If participants from multiple organizations are involved, the planning and coordination will require additional efforts to ensure the participants are integrated effectively into the drill.

If there will be persons enacting the roles of casualties/ victims, aggressors (e.g. active shooter), or portraying the roles of other agencies, who will be playing those roles? In small exercises, it is not unusual to utilize staff from training divisions to play these roles.

The planning of an exercise often starts 6-12 months in advance. The assignment of internal staff may require an extensive amount of staff time to accomplish the planning requirements. If multiple agencies are involved, the added complexity adds to the participation time required from all agencies.

Utilizing Volunteers

Here are some examples of volunteers we have used to play the roles of casualties / victims in various exercises:

- Nursing students

- Paramedic/EMT students

- First aid teams and CERT volunteers

- Firefighting students

- Journalism students (great to play the media or the social media injects)

- Drama students

- Boy/Girl Scouts/Ventures/Rovers

- Army / Air force cadets

- Students from the local military college

- High school classes (requires extensive supervision)

- Volunteer patients from the local medical school (people with medical conditions that are used as practical examples in medical school classes)

- Police auxiliary/reserve officers

- Community volunteer/service organizations (Rotary, Lions, Kiwanis, Kinsmen, etc.), and

- Peers from neighbouring communities (e.g. first responders from other jurisdictions)

Students can be accessed through a couple of different methods; general request to students directly (e.g. college or university students) or going through the school/teachers. In one case we had a college firefighting program coordinator make it mandatory for the firefighting students to participate as casualties.

When using volunteers a number of factors must be considered including:

- **The age of the volunteers.** Youths may require additional supervision or may not be appropriate for certain roles. For example, an age of consent may be required if you are decontaminating "victims".

- **Level of supervision required/available.** For example, a class of high school students may require more supervision than students in a specific program that is closely related to the exercise where the students are more motivated to learn.

- **Their emotional capacity to deal with the stress of the scenario.** Some scenarios may include stressful situations that will not be appropriate for all volunteers.

- **Their physical capacity for the role.** Some roles will require exposure to the weather, being in awkward positions, or being "man-handled" to affect the rescue, and not all volunteers will be capable of filling these roles.

- **The time requirements for the volunteers and those supervising them.**

Personal comfort level and religious restrictions must be taken into consideration in some exercise roles. For example, in a decontamination scenario some people may not be comfortable being in a situation where they have their clothes removed to be scrubbed down, even though they were requested to wear bathing suits under their clothes. Therefore, it is important to ensure the volunteers know what roles are available and what the requirements are for each role so that they understand what they are signing up for.

Other volunteers have few inhibitions. In one situation where we requested a volunteer to play the role of a woman in labour, she poured water over her pants (without our request) to simulate her water breaking while going into labour.

Where we have used the roles of criminals or mentally disturbed individuals in an exercise we always ensure the participants understand and are willing to be restrained by the emergency responders should the scenario go in that direction. This may include handcuffs, leg shackles, hands on restraint, etc.

Although it is done in good sport, there have been a couple of occasions where we have to caution the volunteers from getting too aggressive in these roles as the risk of actual injury increases with a greater resistance from the "patient/criminal" which may result in the first responders becoming more forceful. On a couple of drills we have called "time out" to interrupt a scenario where we were concerned participants were getting too physically involved.

The conditions to which the "casualties/victims" will be exposed must be considered in determining who will play that role. For example, conditions may vary from being inside a climate controlled building to having to lay outside in the weather (rain, snow, high heat).

Depending on the scenario the "victims" may be "trapped" upside down strapped in the seatbelt of an overturned car or floundering in the cold water of a lake or pond. In some exercises debris is piled up on a cement culvert or strong wood frame box constructed to contain a person simulating being buried in the rubble of a building collapse.

Even if you take every measure to minimize the risk, placing people into uncomfortable or potentially hazardous conditions requires that you select the people to play those roles carefully.

It is highly recommended that a liability release is used for all external participants and volunteers. We include an authorization for photography/videography in our release forms and parent/guardian signatures are required for those under 18. Consult your legal services for further advice and form development.

Photo: volunteer registration

CHAPTER 10
PLAYER AND CONTROLLER MANUALS

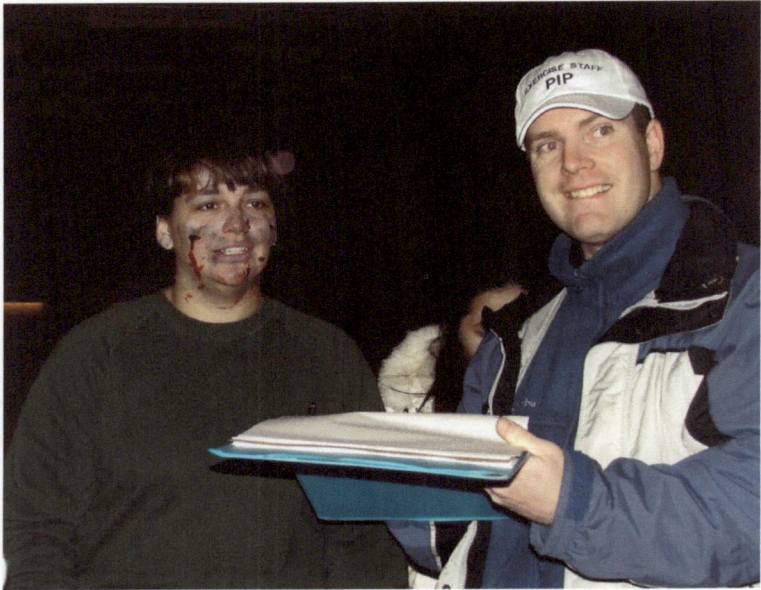

Player and controller manuals are documents that are provided to the participants who are "playing" in the exercise or the controllers who are overseeing the exercise.

These manuals will:

- Outline the key goals and objectives of the exercise. This will help the participants to understand why certain aspects of the emergency plan are being tested while others are not.

- Describe the type and structure of the exercise. For example, the type of tabletop exercise or extent of the functional simulation. This is important so that all participants know to what extent they are expected or allowed, to take the "play".

- The scope and outline of the scenario details will provide the background and set the stage for the incident. For example, the exercise may be simulating a severe storm event while it is a perfectly sunny day.

- Identify the rules of the simulation including the role of the controllers who can start, speed up, slow down, stop, and possibly restart the exercise for various reasons including: exercise direction moving off the objectives, participants being overwhelmed, teachable moments, dealing with real emergencies during the exercise (real emergencies are often referred to as "No Duff"), etc.

- Document the safety issues, precautions and policies specific to the exercise. This will include the fact that everyone is responsible for safety, major health and safety risks, real emergency (No Duff) procedures, as well as the role and authority of the safety officer(s) during the exercise.

- Identify means of communications during the exercise. This will include such things as specific radio channels, phone numbers, etc.

The Controller/Evaluator manuals (which are often combined into one manual) and the Participant manual provide those involved with the exercise key information required to fulfill their roles.

Below is an outline of some of the key information that you may want to include in the manuals. Since the Participant manual is often a smaller version of the Controller/Evaluator manual, with just key areas that are applicable to them, the description below will identify with a (P) any areas that would be in the Participant manual.

By creating the Controller/Evaluator manuals, you can then just copy the sections identified for the Participant manual.

- Introduction – highlights the organization(s) participating in the exercise (P)

- Purpose of the exercise (P)

- Confidentiality clause and reasons for keeping the manual(s) restricted (P)

- Key contacts (planners, lead controller, public information coordinators, etc.)

- Exercise Goals (P)
- Exercise Objectives (broken down by participating agency if it is a multi-agency event) (P)
- Key information on the roles and responsibilities for the various participants and observers. This will include:
  - Planners
  - Lead controller
  - Controllers
  - Evaluators
  - SimCell controllers
  - Safety officer(s)
  - Actors
  - Players (Responders)
  - Observers
  - VIP Observers (may include politicians)
  - Media observers, and
  - Support staff
- Exercise set-up (site, facilities, equipment, vehicles, etc.)
- Participant briefing times/locations/electronic briefings for each of the key groups of participants (P – for players only)
- Safety requirements/rules (P)

- Reporting real emergencies – No Duff (P)

- Site access and security (P)

- Site logistics including parking, food and refreshments, break rooms, etc.

- Controller communications – radio, telephone, texts, etc.

- Player communications – what radio, telephone, email or other communication means they are to use (P)

- Exercise outline – start time, major events, anticipated finish time

- Player guidelines (this section is the participant manual inserted into the controller manual)

- Exercise assumptions (may include the following) (P)

- No fault learning environment

- Response policies and procedures
  - Controllers and evaluators will be well versed in the policies and procedures the players will be implementing
  - Players will respond in accordance with existing plans, policies and procedures
  - Players will react to the information and situations as they are presented, in the same manner as if it had been a real event

- Constructs - Exercise devices to enhance or improve the exercise realism.

- Constraints - Exercise limitations that may distract from the realism. Examples include:

    o Actions or directions may be simulated with no live resource deployment being activated.

    o Some personnel and equipment may be prepositioned at exercise locations.

    o Actors will be playing the role of injured, ill, or otherwise impacted persons. They will receive some orientation and familiarization with their "condition".

    o Controllers may be providing injects, reading scripts, or providing materials.

- Scenario realism

    o Sometimes parts of the scenario may seem implausible. Recognize that the exercise has objectives to satisfy that may require the incorporation of unrealistic aspects.

- Actions participants should take before the exercise. Examples include:

    o Review emergency plans, procedures and any exercise documentation

    o Attend scheduled briefing(s)

    o Wear appropriate uniform, identification, etc. to the exercise

- Actions participants should take during the exercise. Examples include:

  o Players should sign in and take their appropriate roles

  o Wear PPE (personal protective equipment) as appropriate

  o Wear ID vests as appropriate

  o Use the information provided by the controllers to make decisions and take action. Other information may be requested via normal means or simulated means (e.g. calling into the SimCell).

  o Focus on the exercise, not socializing

  o Questions, concerns and clarification of issues should be directed to a controller

  o Communications to outside agencies may be directed to the SimCell

  o Verbalize when taking action which will allow the evaluators to be aware of the actions you are initiating

  o Maintain individual or group logs of your activities

  o If a real emergency occurs the term "No Duff" will be used to notify all persons that a real situation is occurring and the controllers will assess and make decisions on the continuation, pausing, or cancelling of the exercise

- Post Exercise Actions – Examples include:
  - Debriefing location, date, time
  - Participant feedback form
- Exercise scenario (Controller/Evaluator manual only)
  - Scenario background
  - Incident occurrence
  - Activation of response
  - Timeline
  - Injects and inputs
- Evaluation and Post Exercise Activities
- Participant De-briefing (hot wash)
- Controller/Evaluator Debriefing
- Review of evaluation forms
- Post Exercise Summary Report
- Recommendations and Action Plan

As you can see, the Controller/Evaluator and Participant Manuals can become large documents.

Each organization will want to assess the need to provide the information.

CHAPTER 11
MARKETING YOUR EXERCISE

It may sound strange to talk about marketing an exercise, however, we encourage our clients to consider the external benefits that can be gained through the exercise.

An exercise is a great opportunity for the public relations of the organization. The media are often looking for interesting stories and emergency drills can fill that need.

Further, an exercise can be a good news story highlighting preparedness initiatives for clients and stakeholders. Your organization can use the story to encourage the target audience to be better prepared themselves.

The organizations and agencies participating in disaster drills are often publicly funded at least to some degree, meaning they rely on the goodwill of the public and politicians to maintain their level of funding. It is often beneficial to let the public and politicians know about the exercise and the benefits of the event, including being better prepared to serve the public.

Creating a positive impression and goodwill can result in positive consideration in future budget requests, provide reassurance to the community that you are prepared to respond, and create a sense of pride within your responders.

A public relations (PR) plan should be developed so that it complements, but does not interfere with the goals and objectives of the drill. The PR plan is more than just an invitation for the media to attend the exercise, it should have a set of goals and objectives of its own to identify the outcomes of the plan. The messages to be communicated and the methods to communicate them can then be developed.

Consider the following opportunities to market the exercise:

- VIP tours for selected politicians
- Media tours and photo opportunities
- Posting updates and photos on social media
- Ensuring signage and logos are highly visible within your EOC and on-site (e.g. on vehicles, trailers, tents, major equipment and event items like blankets) so that they are captured in photos

When US Airways flight 1549 made an emergency landing on the Hudson River it was an immediate headline. Rescue efforts of NY Waterway ferries, NYPD, FDNY, US Coast Guard, and others were able to ensure the safety of all passengers and crew. A Red Cross responder awaited on shore with boxes of blankets and as each survivor disembarked from the vessels that rescued them, an American Red Cross blanket was wrapped around them. Images were immediately broadcast around the world of the survivors clutching the blankets. In media this would be considered an advertising coup.

Your exercise may receive media attention from the traditional media as well social media. Having your logo and organization name highly visible may result in great publicity for your group.

Media tours and interviews should be scheduled so that the reporters can get a view of the action without interfering with or interrupting the activities. Having a senior person in the organization available to meet the media is important, both to communicate the messages in a personal manner, and to answer questions that they may have. It is important that the person(s) working with the media have training and experience providing media interviews.

Social media should be a vital component of the PR plan but it also needs to be well thought out and planned.

Possible considerations for your social media PR plan may include:

- Inviting well known bloggers as part of the media tour

- Tweeting activities from the exercise as it occurs

- Posting photos and various messages from the exercise as it takes place

- Responding to posts and questions about the exercise in real time

Further, the exercise can be used to gain a connection with the politicians who oversee and fund the programs. By providing a VIP tour the politicians can be shown the benefit of their investment in the emergency management program and the exercises. They can also be provided an opportunity to meet with the media to publicly accept recognition for the success of your program.

We have, in the past, incorporated injects for politicians during their tours to participate in the exercise. For example, during the VIP tour we provided the mayor with a script to interact with the EOC Director. Mayors are usually thrilled to have the opportunity to participate while creating a realistic challenge for the EOC Director to respond to a political request for information.

SAFETY
OFFICER

There is always a risk, although usually a very small one, that a real emergency can occur within or at the same time as the exercise, and it needs to be planned for.

The first step in prevention/mitigation of a true emergency during the exercise is to conduct a HIRA (hazard identification and risk analysis) of the exercise.

Take time to reflect with your exercise planning team on the potential risks that could occur during the drill? Examples may include: participant injury/illness, rescue or heavy equipment malfunction, bystander involvement, severe weather, etc.

As part of the Incident Management System (IMS) being used to plan and implement the drill, a Safety Officer should be appointed to oversee the HIRA process, implement prevention and mitigation strategies, and monitor the activities of the participants during the exercise to ensure that they are safe.

This role will vary in the intensity depending on the size and complexity of the exercise. A tabletop exercise will have a much lower risk than a fully functional exercise utilizing heavy equipment and vehicles.

For functional exercises that take place outside, you must consider the potential weather conditions on the day of the exercise. For example, evacuating seniors from a retirement home will have additional risks based on poor or extreme weather.

No Duff

If a real emergency is detected that requires the attention of the participants, a pre-determined term should be used to alert the participants that the message is not part of the exercise. "**NO DUFF**" is a term that is used internationally for this purpose.

A message may then sound like: "NO DUFF, NO DUFF, NO DUFF firefighter down at the training tower" or "NO DUFF, NO DUFF, NO DUFF, the following resources are required to respond to a tanker truck rollover at ...".

This ensures that the message does not get confused as an exercise input. When the phrase "NO DUFF" is heard, all exercise activity must come to an immediate halt until the situation is identified, assessed, and responded to. At that time the controller(s) will decide whether or not to continue from the point where the exercise was stopped.

The actual term "No DuF" comes from the Royal Air Force (Great Britain) in World War II as a term to mean "No Direction Finding" used before a radio transmission.

In WWII, the location of an aircraft was achieved by using radio towers on the ground to detect and triangulate a radio transmission from the aircraft. To differentiate a real message from a radio transmission for location (a Direction Finding or DF message), it was preceded by the term "No DF", or "No Duff". This quickly came into use by military organizations around the world.

"No Duff" is now a common term used in disaster exercises to confirm "this is a real situation/emergency".

Advising the Public

In a real emergency people will go above and beyond, and even put themselves at risk in order to rescue others. By letting people know that it is a drill, they are less likely to take risky actions. Further, those on the outside looking in are less likely to intervene.

Exercises in areas open to public observation may result in calls to 9-1-1, the media, and social media posts.

As a paramedic student I was a participant in an exercise at an arena that involved EMS, Fire, Police and numerous students in the role of patients. Special effects including smoke generators causing smoke to billow from doorways, responding resources racing from their stations with lights and sirens, and students made up with casualty simulation makeup made it look realistic to the public driving by. One gentleman pulled his car up to the scene, grabbed two college students with moulage who appeared injured, put them in his car and raced off to the hospital a block away.

In one US city, the police were conducting an "active shooter" exercise at a high school where classes were in session. Although the principal knew of the exercise he did not inform the teachers, students or parents that the exercise was going to take place. Students, hearing the lock down message and seeing heavily armed police in the halls, immediately started to use social media to post the "active shooter" situation to family, friends and the greater community. Within minutes panicked parents started arriving at the school along with the media.

In another exercise, a security organization was conducting a hostage scenario which started outdoors and quickly moved into a building, however, they failed to notify the local police department. A bystander, seeing the activity of people armed and in tactical gear, called 9-1-1. A large police response descended on the scene with guns drawn.

Strategies to reduce the risk of confusion include:

- Notifying local emergency services and 9-1-1 call centers that the exercise is occurring

- Notifying persons living or working in the area of the exercise

- Using signage to notify bystanders

- Notifying the media of the exercise

- Attaching the message "this is an exercise" to radio conversations that may be monitored by scanner

- Having a prepared public relations plan

Photo: Advising the public of an exercise

CHAPTER 13
EXERCISE EVALUATIONS

There are several components to an exercise evaluation.

- Impartial Evaluators
- Facilitated Debriefing/"Hot wash"
- Survey/Questionnaires
- Assessing the Objectives
- Lessons learned
- Planning future exercises

Impartial Evaluators

Evaluators are subject matter experts who can provide a knowledgeable and useful assessment of the exercise, identifying the drill's strengths, weaknesses, and opportunities for future improvements. Evaluators can be peers from other jurisdictions, graduate students from universities or exercise experts/consultants.

Evaluators should be provided with an evaluator's manual outlining the goals, objectives, and outline of the exercise so they can evaluate the actions they observe. The evaluators should be assigned to key observation points and keep detailed notes of their observations. These notes should be summed up into a few key points at the end of the drill, and both the key points and the notes submitted to the lead controller.

In order to gain the greatest benefit from the exercise, an evaluation process should be undertaken in order to:

Evaluate the Exercise Itself

Using the goals and objectives, the exercise is evaluated to see if it accomplished what it was initially designed to do. This is a focus on exercise design, planning and implementation.

Evaluate the Response

The most important evaluation is to review the response to the simulation. The exercise is normally not used to evaluate a specific person(s) performance unless it is part of their job performance evaluation. The response is evaluated to identify both strengths and areas for improvement to enhance the response in an actual event. Common evaluation areas include: policies, procedures, equipment, etc. and avoid focus on specific persons.

Evaluations can take different formats. Often there is a participant debriefing / "hot wash" immediately following the exercise. This is a facilitated discussion that should avoid specific identification of people.

It is recommended that participants be given an opportunity to provide their input in a confidential manner, such as a questionnaire, as not everyone is open to discussing issues in a public forum. This also gives opportunity for people to submit more detailed comments and ideas.

A debriefing of the controllers and evaluators will often be held separately.

Facilitated Debriefing/Hot Wash

Following the exercise it is common to have a debriefing or hot wash. This is an opportunity for the exercise participants to speak about their experiences in the drill, challenges they encountered, potential solutions etc. The debriefing should be focused on the response, not individual actions.

A facilitator is useful to ensure that everyone has an opportunity to contribute to the discussion and that no individual monopolizes the discussion. It is useful to review the goals and objectives prior to the hot wash to keep the discussions focused.

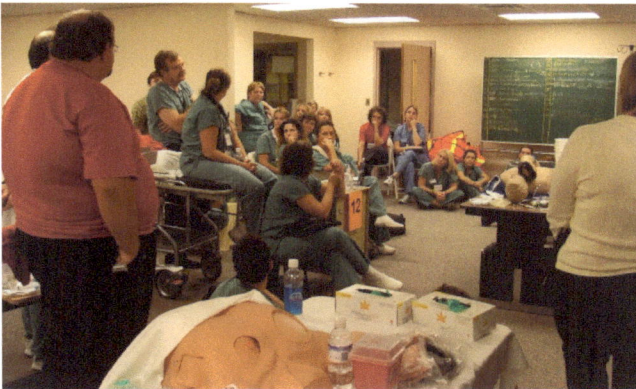

Survey/Questionnaires

Providing the participants with questionnaires, either in a paper format or through an electronic means, will provide them an opportunity to provide input in a very detailed and anonymous way.

Objectives Assessment

One of the first evaluations of the exercise is to determine if the objectives were met. If not, try to determine if the exercise itself was the contributing factor (did not lead to the objectives being addressed) or if the cause was inappropriate procedure/process, lack of training, equipment that didn't meet expectations, etc. For every challenge identified and lesson learned, a task to further examine or rectify the issue should be developed and delegated with a timeline.

Lessons Learned

Once the debriefings are completed, the participant surveys have been gathered, and the controllers/evaluators have had an opportunity to discuss the exercise, the next step is to identify the lessons learned.

The lessons learned will include recommendations for further action such as procedural changes, additional training, equipment and resources to be added, etc. An implementation strategy should accompany the recommendations along with identifying the persons or organizations responsible for implementation of the recommendations.

To best ensure the implementation of the recommendations a timeline, including follow up meetings and deadlines, should be identified.

Too often we have seen good lessons identified in an exercise ignored or put aside until our return visit one year later. Therefore, it is critical to identify the person responsible for following through with action steps and provide a timeline for implementation.

Planning Future Exercises

In order to continue program development and strengthen response to emergencies, the planning of future exercises will often begin immediately following the current exercise. Some communities have repeated a similar exercise, with increased complexity, in subsequent years. It is important to develop scenarios which will create opportunities to ensure that the lessons learned from the previous drill have been implemented and to strengthen the team skillsets. This is particularly useful when there is a specific high risk threat (e.g. tornado alley). At other times, the exercise may be designed around other risks to build on the flexibility of the team.

Have Fun!

An exercise can be very stressful. Participants may feel like they are being tested and under scrutiny. Some participants may not have experience in an emergency situation and the scenario may create anxiety. It is important therefore to break some of the tension and make the event an enjoyable and rewarding experience, much like tackling a game that you want to beat.

While it is a working exercise, your drill should have elements of fun. This will help to ensure that your participants have felt that the exercise was a successful event that they learned from.

Levity will often break some of the stress and tension that can build in a realistic exercise.

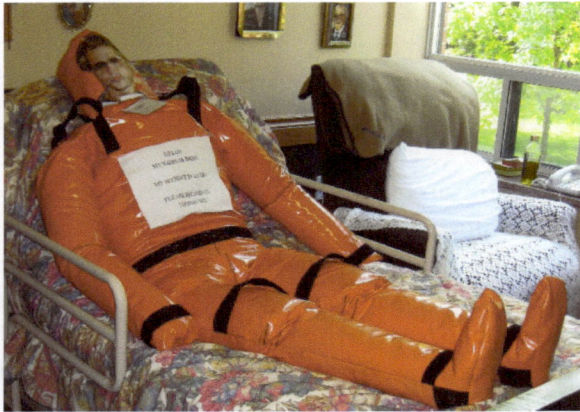

Photo: Nurses at a Long Term Care facility taped a photo of a famous actor's face to a rescue dummy nicknaming it Brad.

CHAPTER 14
LIABILITY AND IMAGE RELEASES

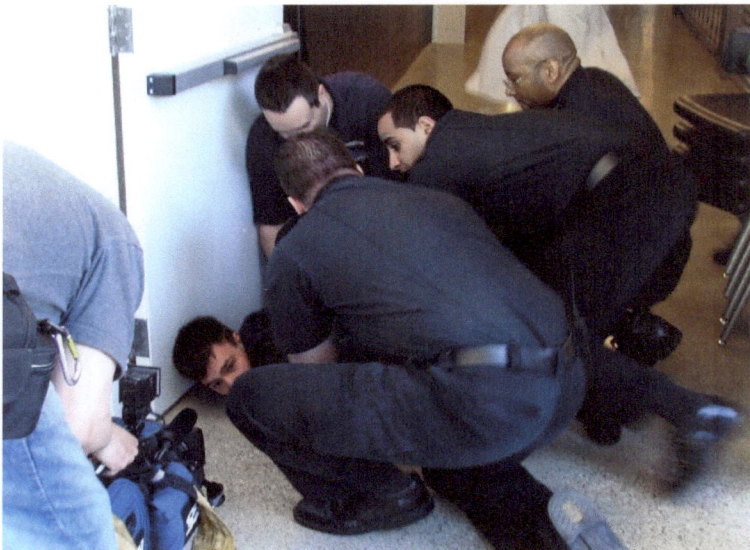

There will be differences in every jurisdiction with regards to the potential liabilities for the organizations planning and participating in the exercise. Therefore, part of the Finance/Administration role will be to ensure insurance, workers compensation, and civil liabilities have been addressed through discussions with solicitors, insurance companies, and government agencies overseeing worker/volunteer safety.

In most instances, there will be a desire to photograph and video (image capture) the exercise for future training, media, publicity, and even legal documentation. Therefore, there will frequently be a Volunteer/Participant Release. This is a document which should be developed or at least reviewed by the organizations' solicitors to ensure the appropriate wording and protection is in place for the organization and planners.

This release may cover the following items:

- Release and waiver of liability
- Identifying that there will be no payment for volunteers
- Identify who is responsible for all arrangements and costs for the participants including travel and meals (e.g. the participant or an organization)
- Agreement that the participants will comply with organizational policies and rules
- Agreement to work/participate under the direction of the controllers/supervisors

- Agreement that the participants will comply with health and safety rules

- Verification of participant health and medical insurance

- Confidentiality clause

- Image release - photography, video, and other media clause

- Workers Compensation Insurance clause

- Parental/Guardian consent (if under the legal age for agreements in the jurisdiction)

- Signature

- Witness signature

- Photo (this is becoming more common to match the person on the contract to the person participating)

- ID check (verification of match between driver's licence or other government ID and participant)

Epilogue

Fast forward ten months and that same team in Texas is facing another tornado threat. This time the tornado leaves far more destruction throughout the city, hitting a local university and leaving so much debris that many roads are inaccessible and power lines are down.

Having taken to heart the lessons learned from their exercise the previous year the team works efficiently to respond to the situation at hand.

Yes, this was another exercise, but this time the exercise was overshadowed by the real storms which were ripping through their area that same week, causing real destruction and chaos. As the exercise concluded, the television screens in the EOC were showing real time images of a series of storms just north of their location. This was a sobering reminder of the importance of their preparedness exercises.

Imagine if every city and organization would take their preparation as seriously!

Emergency Management & Training Inc.

Darryl Culley is the President of Emergency Management & Training Inc., an international consulting firm providing emergency management consulting to governments and organizations of all types and sizes.

Services include:

- Hazard identification and risk analysis
- Prevention and mitigation strategies
- Policy and procedure development
- Response plans
- Business continuity plans / Continuity of operations
- Crisis communications
- Training
- Exercises of all types and sizes
- Post-incident reviews of actual emergencies/disasters

Emergency Management & Training Inc. also provides consulting services to Fire Departments, Emergency Medical Services, and 9-1-1 Telecommunications Centers.

These services include:

- Master Fire Plans
- Strategic planning
- Operational reviews
- Accreditation assistance
- Leadership training

Our team includes professionals with various expertise, experience, and strengths. We bring together the expert team to meet the project needs of each client.

Emergency Management & Training Inc. has been referred to as the "the team in black" in reference to the movie *Men in Black* as we work in the background, get outstanding results, and let the client take the compliments for success.

For more information visit us at: www.emergencymgt.com or email: info@emergencymgt.com

Emergency
Management &
Training Inc.